Publishing

C000065537

2013/14 Edition

OCR
A2 Biology
Unit 4
Revision Workbook

revision
academy

Samantha Richardson

The Revision Academy

The company provides specialised revision that is specific to each exam within the following exam boards:
- AQA
- EDEXCEL
- OCR
- WJEC

We pride ourselves on attention to detail when it comes to revising, and ensure that different learning styles are catered for.
The company currently delivers the following to help boost candidates grades:
- Workbooks specific to each individual exam
- One to one private tutoring sessions
- Intensive small group revision days specific to particular exams

For more information on these services visit our website at www.therevisionacademy.co.uk.

Samantha Richardson, a fully qualified Teacher, started the company in 2008. She has had many successes with her pupils since then and the company has grown steadily. Samantha has a Masters Degree from the University of Southampton and a Post Graduate Certificate of Education from the University of Cambridge. She has worked in both state and private schools and held the position of Head of Science before founding The Revision Academy.

Published by:
Synthus
29 Hickory Lane
Almondsbury
Bristol
BS32 4FR
UK   www.synthus.co.uk/publishing

This workbook has been written specifically to support students preparing for the OCR Specification A2 Biology F214 examination. The content has been neither approved nor endorsed by OCR and remains the sole responsibility of the author.

About this book

Th book contains many different revision techniques to help support you when preparing for the unit 4 exam.

The first section aims to check and test all of your basic knowledge of each topic. Here you should try to include all the key words and terms as they are vital to attaining full marks in exam questions.

The second section gives you space to keep all of your key notes on the most difficult topics in the module.

The third section highlights which subjects have been included in past papers so you may easily access questions on particular topics and give you a possible feel for what may come up in your exam.

The fourth section contains longer questions that require knowledge of several topics and how they are linked together.

The fifth section gives you examples of flow diagrams of the key processes in the module.

The sixth section gives a list of the key concepts in the module that you can use to complete memory spider diagrams to check your current knowledge.

Finally the answers to all the basic questions and linked topic questions are included.

Contents

Basic Questions

This section contains questions on the basic knowledge you will need to know for each topic of the exam.

These questions are useful to answer a few times through your revision. We recommend you do them the first time using your text books and notes to help you, and then answer them a second time using less help.

Homeostasis

These questions are based on how organisms maintain their internal environments and cope with pressures from external sources.

1. What conditions do **enzymes** need to function?

....

....

....

2. What is the difference between a **stimulus** and a **response**?

....

....

....

3. What are the waste products that build up in the body?

....

....

4. What different internal and external environmental stimuli are organisms subjected to?

....

....

....

5. What are the attributes of a good communication system?

....

....

....

6. Describe the difference between the two major systems of communication in cell signalling.

....

....

....

....

7. Give the definition of **homeostasis**.

....

....

8. What functions does **homeostasis** control?

....

....

....

9. How does negative feedback help to maintain the internal environment of an organism?

....

....

....

10. What structures does **negative feedback** require?

....

....

....

11. When would **positive feedback** be used instead of **negative feedback**?

....

....

....

12. What are **endotherms** and **ecotherms**?

....

....

....

13. What are the advantages and disadvantages of being an **ecotherm**?

....

....

....

14. How can **ecotherms** regulate their body temperature?

....

....

....

15. What are the advantages and disadvantages of **endotherms**?

....

....

....

16. Describe the **negative feedback** system for regulating body temperature in **endotherms**.

....

....

....

....

....

The Nervous System

These questions are based on how organisms transmit information around the body in the form of electrical signals.

1. Describe the 6 sensory receptors.

....

....

....

....

2. What does a sensory receptor do?

....

....

....

....

3. How do the proteins in neurone membranes differ from normal cell membrane proteins?

....

....

....

4. What is the purpose of carrier proteins in neurone membranes?

....

....

....

5. What is meant when a membrane is said to be **polarised**?

....

....

....

6. How can a nerve impulse be created?

....

....

....

....

....

7. What is **depolarisation**?

....

....

....

8. What is a **generator potential**?

....

....

....

9. Define the term **action potential**.

....

....

....

10. What are the three types of neurone and what are their functions?

....

....

....

....

....

11. How have neurones become specialised to carry out their functions?

....

....

....

....

12. What is a **resting potential** and how is it maintained?

....

....

....

....

....

13. What is the function of the voltage gated channels?

....

....

....

....

14. What is the **threshold potential** and what happens if it is not reached?

....

....

....

....

....

15. What is meant by the **all or nothing rule**?

....

....

....

16. What is the **refractory period**?

....

....

17. What is a **local current**?

....

....

....

18. How do **local currents** affect the voltage-gated sodium channels?

....

....

....

19. What is the function of the **myelin sheath**?

....

....

....

20. What is **Saltatory conduction**?

....

....

....

21. Describe the structure of a **cholinergic synapse**.

....

....

....

....

....

22. What is a **neurotransmitter**?

....

....

....

23. Why does the **synaptic knob** contain **mitochondria** and **smooth endoplasmicreticulum**?

....

....

....

24. What is the role of the **post-synaptic membrane**?

....

....

....

25. Describe how the impulse is transmitted across the **synapse**.

....

....

....

....

26. What is **acetylcholinesterase** and what is its role in transmission?

....

....

....

....

27. What does the term **summation** mean?

....

....

28. Why are **synapses** so important in the nervous system?

....

....

....

....

29. How does the brain determine how intense **stimuli** are?

....

....

....

....

30. What are the advantages of **myelinated** neurones vs. **non-myelinated**?

....

....

....

....

The Endocrine System

1. What is a **hormone**?

....
....
....

2. What is the difference between an **endocrine** gland and an **exocrine** gland?

....
....
....
....

3. What role do **target cells** play?

....
....
....

4. How is signalling with **hormones** different to signalling with electrical impulses?

....
....
....
....

5. How is the **endocrine** system able to be specific in its functions?

....
....
....

6. What are the two types of **hormone** and how do they differ in the way they work?

....

....

....

7. How does **adrenaline** have an effect on a cell if it cannot enter it?

....

....

....

8. What is the role of the **first messenger**?

....

....

9. What do the **adrenal** glands do?

....

....

....

....

10. What do the cells of the **pancreas** do?

....

....

....

....

11. What is the composition of the fluid that travels through the pancreatic duct?

....

....

12. What does the **Islet of Langerhans** contain and what does it do?

....

....

....

....

13. Describe what happens if blood glucose rises too high.

....

....

....

....

14. Describe what happens if blood glucose drops too low.

....

....

....

....

15. What is the difference between **insulin** and **glucagon**?

....

....

....

16. What is the difference between **diabetes mellitus** and **hyperglycaemia**?

....

....

....

17. Describe the process by which **insulin** secretion is controlled.

....

....

....

....

....

18. What is the difference between **type I diabetes** and type **II diabetes**?

....

....

....

....

19. How can **diabetes** be treated?

....

....

....

....

20. What is **cell metabolism**?

....

....

....

21. What is **myogenic** muscle?

....

....

22. How are the **medulla oblongata** and the **pacemaker** connected?

....

....

revision academy

23. What is the **cardiovascular centre**?

....

....

....

24. How is the heart rate controlled?

....

....

....

....

25. What factors affect the heart rate?

....

....

....

....

....

Excretion

These questions are based on the liver & kidney, and how these organs remove waste from organisms.

1. What is **excretion** and how is it different from **egestion**?

   ....

   ....

   ....

2. What are the main waste products of the body?

   ....

   ....

   ....

3. Where do the waste products come from?

   ....

   ....

   ....

4. What is **deamination**?

   ....

   ....

   ....

5. What organs are involved in **excretion**?

   ....

   ....

   ....

6. Why must carbon dioxide be removed from the body?

   ....

7. Why do nitrogen based compounds need to be removed from the body?

....

....

....

8. What is the name given to **liver cells** and what are some of their functions?

....

....

....

....

9. How does the **hepatic portal vein** differ from normal vessels?

....

....

....

10. How does the **liver** receive blood?

....

....

....

11. How has the structure of the **liver** adapted to ensure its functions are efficient?

....

....

....

12. How are the **liver cells** adapted to their function?

....

....

....

13. What do **kupffer cells** do?

....

....

....

....

....

....

14. What does the **liver** do?

....

....

....

15. What is **urea**?

....

....

....

16. What does the **ornithine cycle** do?

....

....

....

17. How is **urea** formed?

....

....

....

18. How is alcohol detoxification carried out by the **liver**?

....

....

19. What are the benefits of alcohol detoxification?

....

....

....

20. Describe the structure of the **kidney**.

....

....

....

....

....

21. What are the components of a **nephron**?

....

....

....

22. How is **ultrafiltration** achieved in the **nephron**?

....

....

....

....

....

23. What is the name given to the fluid in the **nephron**?

....

24. How does the composition of the fluid in the **nephron** change?

....

....

....

25. What is the role of the **podocytes**?

....

....

....

26. Which membrane filters out the larger substances in **ultrafiltration**?

....

27. What is left in the **capillary** after **ultrafiltration**?

....

....

....

....

28. Describe the process of **selective reabsorption.**

....

....

....

29. What is the role of the following in **selective reabsorption**:

Microvilli –

....

....

....

Co-transporter proteins –

....

....

....

....

Facilitated diffusion –

....

....

....

Sodium-potassium pumps –

....

....

....

30. What is the significance of the **cytochrome p450 enzymes**?

....

....

....

....

31. Describe the process that occurs in the **Loop of Henle**.

....

....

....

....

....

....

....

....

....

....

....

32. How is water **osmoregulated** within the **collecting duct**?

....

....

....

....

....

33. What are the two treatment processes for **kidney** failure?

....

....

....

....

....

34. How can we test for **pregnancy**?

....

....

....

....

....

35. How can we test for **anabolic steroids**?

....

....

....

....

Photosynthesis

These questions are based on how plants produce their own energy using energy from the sun.

1. What is **photosynthesis**?

....

....

....

2. What is the difference between an **autotroph** and a **heterotroph**?

....

....

....

3. What are the products of **photosynthesis** used for?

....

....

....

4. Write the symbol equation for **photosynthesis**.

5. What organelle is used in **photosynthesis**?

....

....

6. Describe the structure of a **chloroplast.**

....

....

....

....

....

7. What is the difference between the **stroma** and the **granum**?

....

....

....

....

8. Complete the table to show the benefits of each adaptation within the chloroplasts.

Adaptation	Benefit
Inner membrane	
Many Grana	
Photosynthetic pigments	
Embedded proteins	
Fluid filled stroma	

revision
academy

9. What is a **photosynthetic pigment**?

....

....

10. How are **accessory pigments** different to **chlorophyll pigments**?

....

....

....

....

11. Why are there two different **chlorophyll pigments**?

....

....

....

....

12. Where does the **light dependent** stage take place?

....

....

....

....

13. What happens to water in the **light dependent** stage?

....

....

....

....

14. What is oxygen used for after it has been created in the **light dependent** stage?

....

....

....

....

15. Why is water important in the **light dependent** stage?

....

....

....

....

16. What **coenzymes** are used in this stage?

....

....

....

....

17. What is the role of the **coenzymes**?

....

....

....

....

18. What is **photophosphorylation**?

....

....

....

....

19. What is an **electron carrier**?

....
....
....
....

20. What is an **electron acceptor**?

....
....
....
....

21. Where does **photophosphorylation** take place?

....
....
....
....

22. What role does light play in the **light dependent** stage?

....
....
....
....

23. What happend to the electrons obtained from water during the **light dependent** stage?

....
....
....
....

24. What is the energy from the electrons used for during the **light dependent** stage?

....

....

....

....

25. How is **ATP** synthesised during the **light dependent** stage?

....

....

....

....

26. What is the difference between **cyclic** and **non-cyclic photophosphorylation**?

....

....

....

....

27. Which photo systems are used within **cyclic** and **non-cyclic photophosphorylation**?

....

....

....

....

28. What is the role of **NADP reductase**?

....

....

....

....

29. Give a definition of the **light independent** stage.

....

....

....

....

30. Where does **light independent** stage take place?

....

....

....

....

31. What is the name of the cycle that occurs in the **light independent** stage?

....

....

....

....

32. What does carbon dioxide do during the **light independent** stage?

....

....

....

....

33. How are the products from the light dependent stage used in the **light independent** stage?

....

....

....

....

34. What is the 5-carbon compound present in the **light independent** stage and what happens to it?

....

....

....

....

35. How does the number of carbons in the compounds of the **Calvin cycle** change?

....

....

....

....

36. What products can **glycerate-3-phosphate** produce?

....

....

....

....

37. What products can **triose phosphate** produce?

....

....

....

....

38. How are **lipids** made?

....

....

....

39. How is the 5-carbon compound reformed?

....

....

....

....

40. What can be produced from glucose?

....

....

....

....

41. Give the full names and abbreviations of all the chemicals found in the **Calvin cycle.**

....

....

....

....

42. What is a **limiting factor**?

....

....

....

....

43. What are the different **limiting factors**?

....

....

....

....

44. How has the concentration of carbon changed within the Earth's atmosphere?

....

....

....

....

45. Where is most of the carbon produced by humans absorbed?

....

....

....

....

46. How is the rate of **photosynthesis** affected by limiting light intensity?

....

....

....

....

47. What are the effects of light on **photosynthesis**?

....

....

....

....

48. How does temperature affect the rate of **photosynthesis**?

....

....

....

....

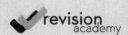

49. What is a **photosynthometer**?

....
....
....
....

50. What can be measured in **photosynthesis** reactions?

....
....
....
....

51. Describe how the effects of light intensity can be investigated on **photosynthesis**.

....
....
....
....

52. Describe how the effects of temperature or carbon dioxide on **photosynthesis** can be investigated.

....
....
....
....

53. What will having more light energy do to the **light dependent** reaction?

....
....
....
....

54. How does this in turn effect the **light independent** reaction?

....

....

....

....

55. What happens when there is less light?

....

....

....

....

56. What does increasing carbon dioxide levels rely on?

....

....

....

....

57. What is the result of increasing carbon dioxide in the **light independent** reaction?

....

....

....

....

58. How does **stomatal** opening effect carbon dioxide uptake?

....

....

....

....

59. What happens if carbon dioxide levels are reduced?

....

....

....

....

60. Why does changing temperature have more effect on the **light independent** reaction than the **light dependent** reaction?

....

....

....

....

61. When can **photorespiration** exceed **photosynthesis**?

....

....

....

....

62. What effect does **photorespiration** exceeding **photosynthesis** have?

....

....

....

....

63. What other effects can increasing temperature have on **photosynthesis**?

....

....

....

....

Respiration

These questions are based on how organisms create energy from different substrates.

1. What is the definition of energy?

....

....

....

....

2. Why is **ATP** used as an energy source?

....

....

....

....

3. What is the difference between **anabolic** and **catabolic** reactions?

....

....

....

....

4. Describe the structure of an **ATP** molecule.

....

....

....

....

5. Why is **ATP** so useful?

....

....

....

6. What processes need energy?

....

....

....

....

7. What is a **coenzyme**?

....

....

....

....

8. Summarise the 4 stages of **respiration**.

....

....

....

....

....

....

....

....

9. What is the role of **coenzymes** in **respiration**?

....

....

....

....

10. What is **NAD** and what does it do?

....

....

....

....

11. How is **coenzyme A** made?

....

....

....

....

12. What is the difference between **NAD** and **NADP**?

....

....

....

....

13. Give a definition of **glycolysis**.

....

....

....

....

14. When something is hydrolysed, what happens to it?

....

....

....

....

15. What happens in stage one of **glycolysis**?

....

....

....

....

16. How is **ATP** used in stage one of **glycolysis**?

....

....

....

....

17. What occurs in stage two of **glycolysis**?

....

....

....

....

18. What is produced as a result of the oxidation of **triose phosphate** in stage three?

....

....

....

....

19. How does **triose phosphate** get converted to **pyruvate**?

....

....

....

....

45

20. Draw a table to show the products of **glycolysis**

21. Where do the following occur:

Where **ATP** is made -

....

....

....

....

Where **ATP** is used -

....

....

....

....

Where **NAD** is reduced -

....

....

....

....

22. What is the function of **mitochondria**?

....

....

....

....

23. Describe the structure of **mitochondria**.

....

....

....

....

24. What determines the distribution of **mitochondria**?

....

....

....

....

25. How are the following adapted to their functions:

Matrix -

....

....

....

....

Outer membrane -

....

....

....

....

Inner membrane -

....

....

....

....

26. What is significant about the use and structure of **ATP synthase** enzymes?

....

....

....

....

27. What is the role of **FAD**?

....

....

....

....

28. Describe the 3 steps of **ATP synthesis** at the **stalked particle**.

....

....

....

....

....

29. What happens during the **link reaction**?

....

....

....

....

....

....

....

30. What enzymes are involved in the **link reaction**?

....

....

....

....

31. What are the products of the **link reaction**?

....

....

....

....

32. What happens to the **reduced NAD** from the **link reaction**?

....

....

....

....

33. What happens during the **Krebs cycle**?

....
....
....
....
....
....
....
....

34. What is the role of **coenzyme A** in the **Krebs cycle**?

....
....
....
....

35 What is the role of **oxaloacetate** in the **mitochondrial matrix**?

....
....
....
....

36. What happens to **citrate** in order for it to be converted to a 5-carbon compound?

....

37. What role does **NAD** have in the conversion of **citrate** to a 5-carbon compound?

....
....
....
....

38. How is the 4-carbon compound created in the **Krebs cycle**?

....

....

....

....

39. How is **oxaloacetate** regenerated and what are the products created?

....

....

....

....

40. Draw a table to show the products made from the **link reaction** and **Krebs cycle.**

41. Why is the enzyme **pyruvate decarboxylase** named so?

....

....

....

....

42. What is **oxidative phosphorylation**?

....

....

....

....

43. Where does **oxidative phosphorylation** occur?

....

....

....

....

44. What is the **electron transport chain**?

....

....

....

....

45. What role do the **coenzymes** play in **oxidative phosphorylation**?

....

....

....

....

46. How are the **H⁺** ions moved across the **inner mitochondrial membrane**?

....
....
....
....

47. What does the movement of ions across the **inner mitochondrial membrane** create?

....
....
....
....

48. What happens to the electrons that were split from the **Hydrogen**?

....
....
....
....

49. What is the role of oxygen in **oxidative phosphorylation**?

....
....
....
....

50. What is the difference between **oxidative phosphorylation** and **chemiosmosis**?

....
....
....
....

51. What are the products of **oxidative phosphorylation** and **chemiosmosis**?

....

....

....

....

52. Why is the potential synthesis of **ATP** molecule numbers rarely reached?

....

....

....

....

53. Why is the pumping of protons not referred to as **active transport**?

....

....

....

....

54. What is Peter Mitchell's theory of **chemiosmosis**?

....

....

....

....

55. What evidence has been found to support the theory?

....

....

....

....

56. What is **anaerobic respiration**?

....

....

....

....

57. In mammals what are the products of **anaerobic respiration**?

....

....

....

....

58. How does the use of **NAD** differ in **anaerobic respiration** to **aerobic respiration**?

....

....

....

....

59. What effect does the presence of **lactate** have on surrounding muscles?

....

....

....

....

60. What could be used to overcome the effects of the presence of **lactate**?

....

....

....

....

61. What enzymes are used during **anaerobic respiration**?

....

....

....

....

62. What happens to the **lactate** that is produced?

....

....

....

....

63. How does **anaerobic respiration** differ in **yeast**?

....

....

....

....

64. What is the role of **NAD** in **alcoholic fermentation**?

....

....

....

....

65. What enzymes are involved in **alcoholic fermentation**?

....

....

....

....

66. What are the products of **alcoholic fermentation**?

....

....

....

....

67. What is a **respiratory substrate**?

....

....

....

....

68. What is a **mole**?

....

....

....

....

69. What allows more **ATP** to be generated?

....

....

....

....

70. What must a compound contain more of to produce more **ATP** in **respiration**?

....

....

....

....

71. What is the theoretical yield of **ATP** from a **carbohydrate** molecule?

....

....

....

....

72. Why is the actual yield different from the theoretical yield?

....

....

....

....

73. What must protein molecules undergo before being used in **respiration**?

....

....

....

....

74. When would **proteins** be used as a **respiratory substrate**?

....

....

....

....

75. Where do the substances enter the **respiration** process?

....

....

....

....

76. Why does **protein** release more energy than **carbohydrate**?

....

....

....

....

77. What are **lipids** converted to in order to be respired?

....

....

....

....

78. Describe the process that fatty acids go through in order to produce energy.

....

....

....

....

79. Which **respiratory substrate** produces the most **ATP** per **mole**?

....

....

....

....

Smart Notes

In this section all of the most difficult parts of the exam are considered and space is made available for you to record the important information.

This can be done in note form, bullet points or diagrams. It is a useful way of keeping all your notes in one place so you may refer back to them easily if you need to.

Synapses

....

....

....

....

....

....

....

....

Synapses

Myelination

....
....
....
....
....
....
....
....

Regulating Insulin

....

....

....

....

....

....

....

....

....

....

....

....

....

....

....

....

Selective Reabsorption

....
....
....
....
....
....
....
....

Loop of Henle

Osmoregulation

....

....

....

....

....

....

....

....

Limiting Factors

....

....

....

....

....

....

....

....

....

....

....

....

....

....

....

....

Importance of Coenzymes

Past Paper Contents

In this section we have reviewed all the available past papers for this particular exam and highlighted the topic of each question within each paper.

This will help you to quickly identify the past paper you wish to use to test particular knowledge of a specific subject.

The past papers can be accessed from the Exam boards website. Alternatively you may register with us and download them from our website www.therevisionacademy.co.uk.

Past Paper Contents

Date	Question	Content
Jan 2010	1	Excretion, respiration equation, diabetes
	2	Glycolysis, anaerobic respiration
	3	Myelin action of neurons, effect of temperature on nerves, synaptic knob
	4	Nephron calculation, dialysis
	5	Photosynthesis, How Science Works question
June 2010	1	ATP, micrograph structure mitochondria, RQ values, endotherms
	2	Micrograph structure liver, ornithine cycle, pregnancy test, steroids
	3	Photosynthesis, photo respiration, O2 concentration
	4	Diabetes, diet
	5	Action potential, MS
Jan 2011	1	Anaerobic respiration
	2	Synaptic knob, nerve gas
	3	Respiration
	4	Osmoregulation, ADH
	5	Adrenaline/Hormone signaling, hormone/nerve control on heart
	6	Calculations, How Science Works question
June 2011	1	Neuron differences, action potentials
	2	Photosynthesis, calculations, How Science Works question
	3	Fatigue & Emphysema, diabetes, irregular heart beat
	4	Pancreas, diabetes
	5	Kidney structure, ultrafiltration, diuretics
	6	Glucocorticoids, photosynthesis & respiration
Jan 2012	1	Homeostasis, insulin (negative feedback), diabetes
	2	Proteins & excretion, pregnancy test, ovulation test
	3	Micrograph structure of chloroplast, DNA & ribosomes in photosynthesis.
	4	Respiration rate, How Science Works, calculations
	5	Micrograph structure of kidney, ultrafiltration, kidney failure
	6	Sensory receptors, neurons

Date	Question	Content
June 2012	1	Communication, neurones, synapses, homeostasis
	2	Kidney, nephrons, loop of Henle
	3	Respiration, electron transport chain, anaerobic respiration
	4	Photosynthesis, limiting factors
	5	Hormeones, endocrine/exocrine glands, insulin
Jan 2013	1	Sensory neurone, resting potential
	2	Liver metabolism, insulin, alcohol
	3	Chloroplast micrograph, light dependent reaction, photosynthesis calculation
	4	Biological terms
	5	ATP - structure & formation
	6	Kidney, ADH

Linked Topic Questions

Increasingly we have seen exam papers include questions that require knowledge across all topics within the unit and also from other previous units.

Therefore we have come up with questions that link topics together and require real in-depth knowledge of the subject content.

These questions will really test your knowledge and your ability to apply it together with other concepts.

Remember to think outside the box, these are not just questions on one particular topic!

1. During starvation certain **respiratory substrates** are no longer available, explain how the body is still able to respire.

....

....

....

....

....

....

....

....

....

....

....

....

....

....

....

....

....

....

....

....

....

....

....

2. What role does the **ornithine cycle** play in **respiration**?

....
....
....
....
....
....
....
....
....
....
....
....
....
....
....
....
....
....
....
....
....
....
....

LINKED TOPIC Q'S

3. How are **excretion** and **respiration** linked?

....

....

....

....

....

....

....

....

....

....

....

....

....

....

....

....

....

....

....

....

....

....

....

....

4. During periods of dehydration, water is retained by the body using the hormone **ADH**. How does this hormone allow the reabsorption of water?

....

....

....

....

....

....

....

....

....

....

....

....

....

....

....

....

....

....

....

....

....

LINKED TOPIC Q'S

5. What is the significance of a cell membrane in hormone secretion?

....
....
....
....
....
....
....
....
....
....
....
....
....
....
....
....
....
....
....
....
....
....
....
....

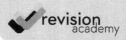

6. How do **insulin** and **glycogen** regulate blood glucose levels?

LINKED TOPIC Q'S

7. Describe the role of **respiration** in temperature regulation of **endotherms**.

....
....
....
....
....
....
....
....
....
....
....
....
....
....
....
....
....
....
....
....
....
....
....
....
....

8. How may a dramatically reduced salt intake effect the sensory system?

....
....
....
....
....
....
....
....
....
....
....
....
....
....
....
....
....
....
....
....
....
....
....
....

LINKED TOPIC Q'S

9. A person consumes extremely high amounts of salt constantly and begins to feel numbness, loss of taste, hearing and other senses. Why is the person experiencing these symptoms?

....
....
....
....
....
....
....
....
....
....
....
....
....
....
....
....
....
....
....
....
....
....

10. MS is a disease which effects the **myelin sheath** of **neurons**. What symptoms may a sufferer experience and why?

....

....

....

....

....

....

....

....

....

....

....

....

....

....

....

....

....

....

....

....

....

LINKED TOPIC Q'S

11. **Cyanide** is a **respiratory inhibitor**, which declines and ultimately ceases the production of **ATP**. How might this effect nervous transmission?

....

....

....

....

....

....

....

....

....

....

....

....

....

....

....

....

....

....

....

....

....

12. Why is it important for **mitochondria** to be present among **neurons**?

....

....

....

....

....

....

....

....

....

....

....

....

....

....

....

....

....

....

....

....

....

LINKED TOPIC Q'S

12. Why is it important for **mitochondria** to be present among **neurons**?

....

13. How might a **calcium** deficiency effect the **sensory system**?

14. A certain gas is known to non-competitively inhibit **acetylcholinesteras**, what would be the effect of inhalation of this gas?

LINKED TOPIC Q'S

15. Some substances are able to bind to the **postsynaptic membrane** in **neurons**, blocking them from receiving any further substrates. What effect will this have on **action potential** transmission?

....

....

....

....

....

....

....

....

....

....

....

....

....

....

....

....

....

....

....

....

16. A person constantly consuming a poor diet, particularly high in sugars, begins to feel diabetic symptoms. Explain the effect of poor diet on **insulin** production.

LINKED TOPIC Q'S

17. During an accident a patient suffers a severed **vagus nerve**. What effects will this have on the body and how could doctors remedy the situation?

....
....
....
....
....
....
....
....
....
....
....
....
....
....
....
....
....
....
....
....
....
....
....

18. Describe the cascading effects of exercise on the body's systems.

....
....
....
....
....
....
....
....
....
....
....
....
....
....
....
....
....
....
....
....
....
....
....
....
....

LINKED TOPIC Q'S

19. Some doctors claim that a small, regular intake of **alcohol** can be beneficial. Explain why this may be so.

....

....

....

....

....

....

....

....

....

....

....

....

....

....

....

....

....

....

....

....

....

....

....

....

Flow Diagrams

This section gives contains flow diagrams of the major processes within the exam.

These can be used to test your knowledge by covering up the flow diagram and trying to remember each step.

In addition, writing the flow diagram out as a paragraph as you would in an exam question can be a really useful revision technique.

This is a useful tool to cement your knowledge on some tricky processes.

Homeostasis

A system will have a desired optimum for a certain condition

↓

Receptors detect any changes from this optimum

↓

If a change occurred the receptor will send instructions to a control system

↓

This control system sends instructions to effectors

↓

These effectors make a correction to return to the optimum

↓

Signals are then fed back to the control system that everything is back to normal

revision academy

FALL IN CORE TEMPERATURE

↓

Hypothalamus detects the change (thermoregulatory centre)

↓

The liver, muscles and skin receive impulses and hormonal signals

↓

The signals tell the liver to respire more, the skin to raise the hairs, the muscles to move spontaneously

↓

More heat is produced and the core temperature rises

Maintaining Body Temperature (Endotherms)

RISE IN CORE TEMPERATURE

↓

Hypothalamus detects the change (thermoregulatory centre)

↓

The liver, muscles and skin receive impulses and hormonal signals

↓

The signals tell the liver to respire less, the sweat glands in the skin to produce sweat, the muscles not to move spontaneously

↓

Less heat is produced and the core temperature falls

Resting:
Na$^+$/K$^+$ pump is on pumping 3Na$^+$ out and 2K$^+$ in – more negative inside cell

↓

Stimulus:
A Stimulus opens the Na$^+$ channels, Na$^+$ move into the neurone

↓

Threshold:
If Na into neurone reach -50mV voltage gated Na$^+$ channels open, major influx of Na$^+$, more positive inside neurone (+40mV)

↓

Repolarise:
+40mV inside triggers Na$^+$ channel to close & K$^+$ channels open. K$^+$ diffuse out & it becomes more negative inside the neurone

↓

Hyperpolarise:
Too many K$^+$ outside closes K$^+$ channel, Na$^+$/K$^+$ pump turns on, back to resting

FLOW DIAGRAMS

revision academy

Synapses

An action potential travels along the neurone and arrives at the synaptic knob, this triggers calcium channels to open, calcium diffuses into the synaptic knob

↓

This triggers the movement of vesicles containing neurotransmitters to fuse with the pre-synaptic membrane and release the neurotransmitter into the synaptic cleft

↓

The neurotransmitter diffuses across the cleft and binds to complimentary receptors on the post synaptic membrane

↓

This triggers the opening of sodium channels in the post synaptic neurone and sodium diffuses into the post synaptic neurone

↓

If this generator potential reaches the threshold level, then an action potential is initiated in the post synaptic neurone

↓

The enzyme complimentary to the neurotransmitter then breaks down the neurotransmitter

revision academy

Steroidal hormones are able to pass through the plasma membrane

↓

Once in the cell the steroid hormone will bind with its complimentary receptor

↓

This then triggers the hormone – receptor complex to enter the nucleus through nuclear pores

↓

The complex acts as a transcription factor during the cells protein synthesis

↓

The transcription factor will turn certain genes on or off

↓

This controls when certain proteins are synthesised

Effects of Peptide Hormones

Peptide hormones are <u>not</u> able to pass through the plasma membrane

↓

Glycoproteins integrated within the cell membrane contain specific receptors on the outside of the cell

↓

Hormones complimentary to these receptors bind to them

↓

This causes a change of the glycoprotein on the inside of the cell

↓

This change activates a second messenger on the inside of the cell

↓

A series of changes may occur and the secondary messenger then acts as a transcription factor in the nucleus

↓

This controls when certain proteins are synthesised

BLOOD GLUCOSE IS TOO HIGH

Beta cells detect the high glucose concentration

Beta cells secrete insulin into the blood

Liver & muscle cells that contain insulin receptors detect the hormone

Glucose channels allow glucose to enter the cells

The glucose is converted to glycogen and stored, or used in respiration or converted to fats

The glucose level in the blood falls

FLOW
DIAGRAMS

revision academy

Regulation of Blood Glucose

BLOOD GLUCOSE IS TOO LOW

↓

Alpha cells detect the low glucose concentration

↓

Alpha cells secrete glucagon into the blood

↓

Liver cells that contain glucagon receptors detect the hormone

↓

The glycogen in the cells is converted to glucose and released into the blood stream

↓

Fatty acids are used in respiration, fats and amino acids are converted into glucose

↓

The glucose level in the blood rises

High levels of glucose in the blood moves into the beta cells

↓

At this point potassium is flowing out of the cell through channels

↓

Glucose is metabolised to produce ATP

↓

The ATP closes the potassium channels and the cell membrane depolarises

↓

This opens the calcium channels and calcium moves in down its concentration gradient

↓

The presence of calcium ions triggers vesicles containing insulin to fuse with the cell membrane

↓

Insulin is then released from the cell

FLOW DIAGRAMS

Control of Heart Rate by Baroreceptors

DECREASE IN BLOOD PRESSURE

The change is detected by carotid baroreceptors

These are located in the common carotid artery

This change triggers an impulse that is sent along a sensory nerve

The impulse arrives at the medulla oblongata

The impulse is sent down the spinal cord to the sympathetic chain ganglia

The impulses passed along the sympathetic nerve to the heart

The effect is that the heart rate increases & causes the blood pressure to increase

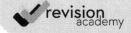

INCREASE IN BLOOD PRESSURE

The change is detected by carotid baroreceptors

These are located in the common carotid artery

This change triggers an impulse that is sent along a sensory nerve

The impulse arrives at the medulla oblongata

This in turn sends an impulse along the parasympathetic nerve

This arrives at the terminal ganglion in the heart

The effect is that the heart rate drops & causes the blood pressure decrease

Excess amino acids are transported to the liver
Amino acids are deaminated into ammonia and a keto acid

The ammonia is then run through the ornithine cycle where carbon dioxide and water is added to produce urea and water

The urea is transported in the blood to the kidneys, arriving through the renal artery

The blood travels into the afferent arteriole where it enters the glomerulus under high hydrostatic pressure, this forces small substances out of the blood into the Bowman's capsule of the nephron in the process known as Ultrafiltration

The Filtrate now travels into the proximal convoluted tubule where Selective Reabsorption occurs. Sodium is actively transported back into the blood, carrying glucose and amino acids with it

A low water potential is created in the blood and water moves out of the nephron via osmosis

The filtrate moves into the loop of Henle. The ascending limb pumps sodium out into the medulla which creates a low water potential in the medulla, water moves out of the descending limb via osmosis

The filtrate get more concentrated as it moves down the loop and sodium moves out by diffusion, until the concentration decreases and moves up the ascending limb

The filtrate moves through the distal convoluted tubule and to the collecting duct where more water can be removed

Osmoreceptors within the hypothalamus detect changes in water potential of the blood

If the water potential is low within the blood, water moves out of the osmoreceptors

The osmoreceptors shrivel up and this stimulates neurosecretory cells in the hypothalamus

Neurosecretory cells produce ADH and it flows down an axon to the posterior pituitary gland and is stored

Once stimulated, the neurosecretory cells send signals down the axon to release ADH into the blood

ADH flows through the circulatory system to the collecting ducts of nephrons

ADH causes pores in the collecting duct to open & water is reabsorbed by the body

Osmoregulation

FLOW DIAGRAMS

Light Dependent Reaction

Photolysis of water occurs where light splits a water molecule into H^+ & OH^-

↓

On the thylakoid membranes the electron from H is accepted by photosystem II

↓

The oxygen is removed as a waste product

↓

Light hits photosystem II and the electron gains energy & arrives at an electron acceptor

↓

The electron then moves along an electron transport chain which allows the synthesis of an ATP molecule

↓

The electron then arrives at photosystem I

↓

Light hits photosystem I and the electron gains energy

Cyclic

Some electrons at this point then travel along the electron transport chain again to produce more ATP

Non- Cyclic

Some electrons move from an electron acceptor and rejoin with H^+

↓

This H is accepted by NADP (now reduced NADP) and move to the light independent reaction

In the stroma of the chloroplasts, ATP and Reduced NADP are used in the CALVIN CYCLE

↓

Carbon dioxide combines with RuBP in the presence of the enzyme RUBISCO

↓

A 6 – carbon compound is made, which is very unstable

↓

This 6 – carbon compound splits into 2 molecules of GP (each a 3 – carbon compound)

↓

GP is reduced using the H from reduced NADP and energy from ATP

↓

A new 3 – carbon compound known as TP is formed

↓

Some TP is then reformed into RuBP using ATP

↓

Some TP is synthesised into glucose

↓

This glucose is used in reparation, used to make amino, lipids etc.

Aerobic Respiration

GLYCOLYSIS:
Glucose is converted to pyruvate through the addition of phosphates and removal of hydrogen

↓

LINK REACTION:
Pyruvate is converted to acetate through dehydrogenation and decorboxylation

↓

Coenzyme A transports the acetate to the Krebs cycle in the form of acetyl coenzyme A

↓

KREBS CYCLE:
Acetate combines with oxaloacetate to form citrate (6C)

↓

Citrate is decarboxylated twice and dehydrogenated 4 times to produce oxaloacetate once more

↓

The waste CO_2 is removed and Hydrogen taken by NAD & FAD to inner mitochondrial membrane

↓

OXIDATIVE PHOSPHORYLATION:
Hydrogen is split into its proton and electron

↓

The protons are pushed across the membrane using energy from the electrons

↓

A proton gradient is formed which flows back across the membrane through a stalked particle and produces ATP

revision academy

NAD & FAD bring hydrogen from glycolysis and the krebs cycle to the inner mitochendrial membrane

↓

NAD & FAD give up hydrogen and are free to accept more hydrogen again

↓

Hydrogen splits into is proton and electron

↓

The electron is accepted by the first electron carrier in the membrane

↓

The energy from this electron is used to push the proton across the membrane into the inter membrane space

↓

The electron moves along the electron transport chain in the membrane, reducing each carrier in turn

↓

The hydrogen is now more concentrated in the intermenbrance space and so flows back to the matrix through the stalked particle

↓

This turns the head of the stalked particle which contains ATPase and so ATP is synthesised

↓

The protons and electrons are now accepted by an oxygen molecule (final electron acceptor)

↓

Water is formed as a waste product and removed

FLOW DIAGRAMS

Protein as a Respiratory Substrate

1. Excess amino acids are deaminated in the liver

↓

Amine group is removed and converted to urea

↓

Rest of molecule is converted to glycogen or fat & then stored

↓

2. If organism is starved/fasting/prolonged exercise

↓

Protein is hydrolysed to amino acids

↓

Some converted into pyruvate/acetate

↓

Some goes straight into krebs and continue respiration as normal

↓

Protein has more H per mole than glucose, so more H available for oxidative phosphorylation, so more energy produced

Oxygen is the final electron acceptor in oxidative phosporylation

↓

If no oxygen, electrons build up in the electron transport chain in the inner mitochondrial membrane

↓

NADH unable to release H as electrons have no where to go

↓

NADH not reoxidised (not able to get rid of H)

↓

No NAD available to dehydrogenate Citrate in Krebs

↓

Krebs cycle cannot continue

Mammalian Anaerobic Respiration

Normal Gylcolysis

↓

Produces Pyruvate & NADH

NAD (reoxidised, used in further glycolysis)

↓

NADH needs to reoxidise but be, as no oxygen. Pyruvate becomes hydrogen acceptor

↓

Pyruvate becomes lactate by accepting H from NADH (aided by lactate dehydrogenase)

↓

Lactate reduces pH (more H from NADH)

↓

Low pH effects enzymes in the metabolic process, this leads to muscle fatigue

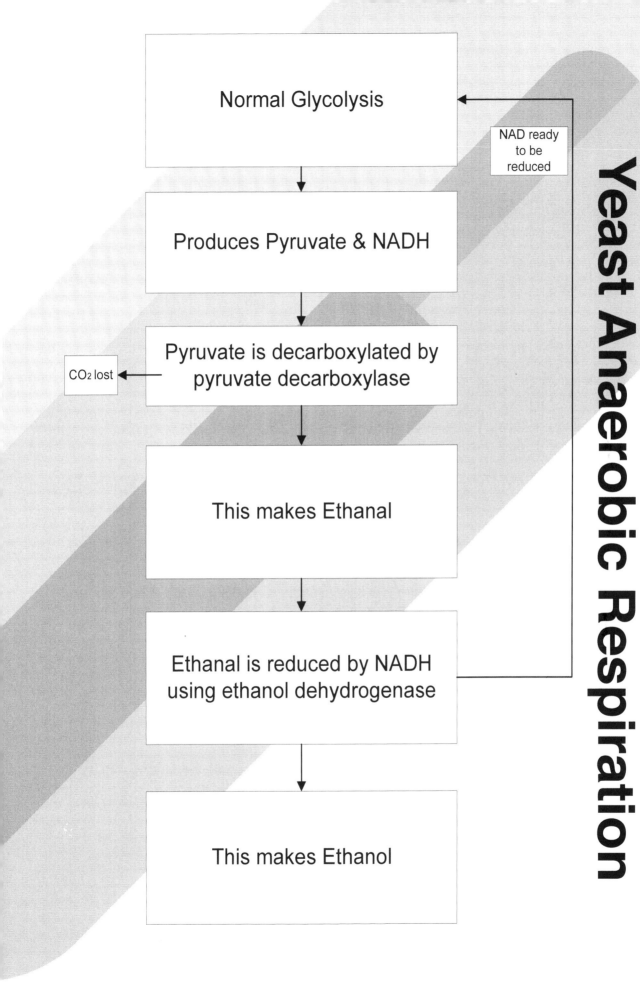

Normal Glycolysis

Produces Pyruvate & NADH

Pyruvate is decarboxylated by pyruvate decarboxylase

CO_2 lost

This makes Ethanal

Ethanal is reduced by NADH using ethanol dehydrogenase

This makes Ethanol

NAD ready to be reduced

Yeast Anaerobic Respiration

FLOW DIAGRAMS

Lipid as a Respiratory Substrate

Triglycerides are broken down by a hydrolysis reaction into fatty acids and glycerol

↓

The glycerol is converted to glucose and respired

↓

Fatty acids combine with a coenzyme A molecule using energy from the ATP to AMP reaction

↓

The fatty acid/coenzyme A move into the mitochondrial matrix

↓

This gets broken down into 2-carbon acetyl groups attached to coenzyme A

↓

This produces reduced NAD & FAD

↓

The acetate groups then enter the krebs cycle

↓

More reduced NAD, FAD & ATP are formed

↓

The reduced NAD & FAD lose their H to the oxidative phosphorylation reaction

↓

More ATP is produced

Memory Spider Diagram Topics

Here you will find a list of the key topics of the module. For each of these topics you can construct memory spider diagrams every couple of weeks to track the progress of your knowledge.

To do this, take a piece of paper and put the title of the topic in the middle of the page in black pen.

Then, without the aid of any text books or your notes, in black pen write down everything you can remember about that topic.

Once you cannot remember anymore, find the chapter concerning that topic in your text books and notes and look up the information that you missed.

In red pen add the missed out material to your spider diagram.

Keep your spider diagram for your records and use it to compare to subsequent diagrams that you construct on the same subject to show the proportion of black to red and see how you are progressing.

- Homeostasis

- Nervous Transmission

- Regulated Blood Glucose

- Control of Heart

- The Liver

- The Kidney

- Photosynthesis

- Limiting Factors

- Aerobic Respiration

- Anaerobic Respiration

- Kidney Failure

Answers

All the answers to the basic questions and linked topic questions can be found in this section.

Homeostasis

1. Optimum pH, optimum temperature, solution, no inhibitors.
2. The stimulus is the change that happens in the environment and the body has a response to it.
3. Carbon dioxide, amino acids.
4. External environments receive changes in temperature, internal environments respond to waste products.
5. Cells are able to send messages and communicate with each other quickly and sustain communication in the short and long term.
6. Hormones release signals into the blood and electrical impulses send signals along neurones.
7. Maintaining the internal environment of an organism even with external changes.
8. Body temperature, blood glucose level, water potential, blood pressure, carbon dioxide concentration.
9. It reverses any changes in conditions and maintains the internal environment.
10. Receptors to sense changes, a system to receive the message and cells to implement a response.
11. To increase any changes, can sometimes be harmful, sometimes beneficial during pregnancy to stimulate uterine contractions.
12. Endotherms can maintain internal body temperature. Ecotherms need external sources to maintain body temperature.
13. The use of food is mainly for growth and they can go without food for long periods. They are not very active in cooler temperatures.
14. Hide in burrows, sit in the sun to absorb heat.
15. They can maintain constant body temperature despite external conditions so can live in colder places. The energy from food is used primarily to maintain body temperature, so more food must be ingested.
16. If there is a rise in body temperature the hypothalamus detects the change and the endocrine and nervous system carry signals to the skin, muscles and liver, these generate less heat and the body temperature falls.
If there is a fall in body temperature the hypothalamus detects the change and sends the signal to the liver, skin and muscles to generate more heat and the body temperature rises.

The Nervous System

1. Rods and cones in the retina of the eye detect light, Olfactory cells lining the nasal cavity detect chemicals, taste buds in the tongue detect chemicals, pressure receptors in the skin detect pressure, Sound receptors in the ear detect vibrations in the air, muscle proprioceptors detect changes in the length of muscles.
2. Detects a change in the environment.
3. They are specific to sodium and potassium ions.
4. Pump ions across the membrane.
5. When there is a potential difference across the membrane.
6. When the permeability of the cell membrane to sodium ions is altered, this causes a move in sodium ions and changes the potential difference across the cell membrane.
7. When the polarisation across the cell membrane is lost and sodium ions enter the cell.
8. When sodium ions enter the cell and cause a small depolarisation.
9. This is an all or nothing response when the membrane gets depolarised to +40mV, more sodium enters and the action potential is reached.
10. Sensory neurones detect stimuli in the environment, Motor neurones transfer the response to the effectors, and relay neurones connect the sensory neurones and the motor neurones together.
11. They have become long and thin to transmit the electrical signal, they contain proteins to allow the passage of sodium and potassium ions to pass through them, they have an insulating sheath known as a myelin sheath.
12. When there is a potential difference across the membrane and sodium is being actively pumped out of the cell.
13. They contain a gate that controls when charged particles can pass through them.
14. When the potential difference across the cell membrane is around -50mV, if this is not achieved then an action potential is not fired.
15. When depolarisation is enough to reach the threshold potential and an action potential is initiated, this will carry on until it reaches the end of the neurone.
16. The time when the cell membrane is recovering from the passage of an action potential and another action potential cannot be initiated so that the action potential only travels in one direction.
17. When ions move along the neurone because of high concentrations when they enter the cell.
18. They cause them to open.
19. Insulates the electrical impulse.
20. The way the action potential moves from each node of ranvier to the next by jumping between them.
21. It is the end of a neurone where there is a gap between it and the next neurone where neurotransmitters diffuse across, carrying the signal to initiate an action potential in the post synaptic membrane.
22. Is a chemical that carries the signal across the synaptic cleft to the post synaptic membrane and initiating a subsequent action potential.
23. Mitochondria to provide energy for active processes and smooth endoplasmic reticulum for storage and supply of calcium ions.
24. Contains receptor sites that are specific to neurotransmitters, when the neurotransmitters bind to the receptor sites the sodium channels open.

25. The action potential arrives at the end of the neurone in the synaptic knob which causes the calcium protein channels to open. As calcium ions move into the synaptic knob this initiates the vesicles to move towards the pre-synaptic membrane and fuse with it. Once fused the vesicles release the neurotransmitter into the synaptic cleft. The neurotransmitters then move by diffusion across the synaptic cleft to reach the postsynaptic membrane of the next neurone. The neurotransmitter binds to the receptors on the post synaptic membrane which opens the sodium channels of the next neurone. Sodium then floods in and begins the next action potential in that neurone. An enzyme then breaks down the neurotransmitter in the synaptic cleft.
26. It is an enzyme that breaks down acetylcholine.
27. When lots of small depolarisations cause a large potential difference across the membrane.
28. They can link several neurones together, the signal will only travel in one direction, summation can occur, acclimatisation can occur, used in memory.
29. Depends on the frequency of action potentials being generated
30. The transmission of the impulse is faster.

The Endocrine System

1. A chemical message released by glands in the endocrine system that travel in the blood and acts upon cells with their complimentary receptors.
2. Endocrine hormones are released directly into the blood and exocrine hormones are secreted through a duct first and then into the blood.
3. They contain the complimentary receptors for hormones.
4. Slower, uses the blood and can have long-lasting effects.
5. Use of specific target cells with complimentary receptors and producing different effects within cells and tissues.
6. Protein hormones attach to receptors outside the cell and do not enter, Steroid hormones may pass into the cell and influence DNA.
7. When it attaches to its complimentary receptor outside the cell this activates the enzyme adenyl cyclase inside the cell, this enzyme converts ATP into cyclic AMP which can then activate other enzymes within the cell.
8. It is the hormone that travels in the blood and attaches to receptors on cells.
9. Adrenal medulla releases adrenalin that prepares the body for activity, Adrenal cortex release steroidal hormones which control the concentration of ions in the blood and metabolism in the liver.
10. Alpha cells secrete glucagon, Beta cells secrete insulin.
11. It contains amylase, tripsinogen, lipase and sodium hydrogencarbonate.
12. Contains the alpha and beta cells that secret hormones.
13. The beta cells detect the glucose and release insulin, the insulin acts on cells in the liver and muscles. It activates cAMP in these cells which opens glucose channels in the cells so glucose moves into theses cells. Here the glucose is converted to glycogen and stored, or the glucose is converted to fats or used in respiration.
14. The alpha cells detect a drop in blood glucose level and release glucagon. This hormone acts on the hepatocytes in the liver, the glycogen in the cells is then converted to glucose and fatty acids are used for respiration.
15. Insulin reduces blood glucose levels, glucagon increases blood glucose levels.
16. Diabetes is when the body cannot control blood glucose levels, hyperglycaemia is when the blood glucose levels are too high.
17. Potassium ion channels of the beta cells are closed when glucose levels are high outside the cells and the glucose moves in and respiration produces ATP. When these channels close it causes an increase in charge inside the cell. This causes calcium channels to open so calcium enters the cells and this causes vesicles of insulin to fuse with the cell membrane and release the hormone into the blood stream by exocytosis.
18. Type 1 – the body can no longer manufacture enough insulin. Type 2 – The body still produces insulin but the cells become less responsive to insulin.
19. Type 1 – insulin injections, type 2- regulation of diet.
20. When chemical reactions occur in the cytoplasm of cells.
21. Muscle that can create its own contractions.
22. Through the vagus and accelerator nerves.
23. Part of the brain in the medulla oblongata where processes and concentrations are monitored and where messages can be sent to the heart in increase or decrease activities.
24. The cardiovascular centre sends impulses to the heart along the vagus nerve.
25. Muscle use, Carbon dioxide concentration, Adrenalin concentration, Blood pressure.

Excretion

1. Excretion is when waste from metabolic processes is removed form the body, egestion is when waste that has not entered cells is removed from the body.
2. Urea and carbon dioxide.
3. Carbon dioxide from respiration and urea from excess amino acids.
4. When amino acids have their amine group removed and ammonia is produced.
5. The Liver and kidneys for urea, and the lungs and heart for carbon dioxide.
6. Haemoglobin carries both oxygen and carbon dioxide so if there is too much carbon dioxide then less oxygen can be transported. Carbon dioxide can form carbonic acid in the blood, which causes many problems in the body.
7. They cannot be stored in the body.
8. Hepatocytes are involved in glucose concentration management, protein synthesis, detoxification, storage of carbohydrates.
9. There are capillaries at both ends of the vein as it carries blood from the digestive system to the liver.
10. Blood from the heart that is oxygenated arrives via the hepatic artery, blood also arrives from the digestive system through the hepatic portal vein, this blood is deoxygenated. Blood leaves via the hepatic vein.
11. The cells are arranged to ensure maximum surface area for blood vessels and the liver cells. Many vessels run through the lobes of the liver.
12. They are not specialised for any function in particular and can carry out a range of functions. They posses microvilli on their surface and are of a cuboid shape.
13. They act like macrophages and break down old red blood cells.
14. Forms urea and detoxifies alcohol.
15. It is formed when amino acids have their amine group removed and carbon dioxide added.
16. Produces urea and water from ammonia and carbon dioxide.
17. Through deamination of amino acids and through the addition of carbon dioxide in the ornithine cycle.
18. Ethanol is broken down by the enzyme ethanol dehydrogenase in the liver to form ethanol. Ethanal is then broken down by ethanal dehydrogenase into ethanoic acid, this combines with coenzyme A to form acetyl coenzyme A which can be used in respiration.
19. Extra respiratory substrate.
20. The renal artery that carries blood to the kidney, renal vein that carries blood away from the kidney in the pelvis of the kidney. Many nephrons sit across the cortex and medulla of the kidney. The capsule surrounds the cortex and the ureter connects the kidney with the bladder.
21. Afferent arteriole, efferent arteriole, glomerulus, bowman's capsule, proximal convoluted tubule, loop of Henle, distal convoluted tubule, collecting duct.
22. Through increased hydrostatic pressure due to narrowing of blood vessels, increased surface area of the Bowman's capsule through podocytes.
23. Filtrate.
24. Proximal convoluted tubule – selective reabsorption decreases the concentration of salts, water and sugars. Loop of Henle – Water is removed and salt concentration increases, then decreases in ascending limb. Collecting duct - water is reabsorbed into the blood so water potential decreases.

25. Increase surface area.
26. Basement membrane.
27. White and red blood cells, larger proteins, some water.
28. Sodium is actively transported into the blood causing a lower water potential within the blood, so water moves via osmosis across the proximal convoluted tubule wall into the blood. The active transport of the sodium also facilitates the diffusion of glucose and some amino acids back into the blood.
29. What is the role of the following in selective reabsorption:
    Microvilli – Increase the surface area for reabsorption.
    Co-transporter proteins –They transport the glucose and amino acids when sodium is actively transported.
    Facilitated diffusion –Move glucose and amino acids out of the nephron.
    Sodium-potassium pumps –Actively removing sodium out of the nephron to create a low water potential within the blood.
30. Breaks down toxic molecules.
31. Counter current multiplier – Sodium is actively pumped into the medulla from the ascending limb which reduces the water potential of the medulla. Water then moves out of the permeable descending limb via osmosis. The filtrate then increases in concentration to its highest at the base of the loop. Ate the base concentration is so great that sodium diffuses out, reducing the concentration of filtrate as it begins to move up the ascending limb.
32. Osmoreceptors within the hypothalamus of the brain detect changes in the water potential of the blood. If the water potential is low, water moves out of the receptors into the blood and the receptors shrivel up causing an electrical impulse to be sent to the neurosecretary cells within the hypothalamus. These then release ADH from the posterior end of the pituitary gland. The ADH travels through the blood and acts on the cells of the collecting duct causing them to become more permeable to water. This allows water to move via osmosis back into the blood stream and be retained by the body to increase blood water potential.
33. Dialysis where a artificial kidney system is used, or kidney transplant where a new kidney is placed into the body.
34. Pregnancy produces the hormone hCG, a pregnancy test can detect the hCG hormone through the use of monoclonal antibodies. The test contain free hCG antibodies that are labelled with a blue dye. The hCG molecule binds to the antibody where it gets carried to immobilised antibodies positioned on a strip. The blue line indicates the hCG is present and therefore pregnancy.
35. Gas chromatography – the urine sample is vaporised into a gas and passed along a tube which contains an agent that can aborb substances. The substances within the urine will all be of different masses and be absorbed at different stages along the tube. The results are compared with known drug chromatograms.

ANSWERS

Photosynthesis

1. Process where light energy from the sun is transformed into chemical energy. This chemical energy is used to make large organic substances from inorganic ones.
2. An autotroph uses light energy to synthesise its organic molecules, where heterotrophs ingest and digest complex organic chemicals and release the energy from them.
3. The products are used in respiration.
4. $6CO_2 + 6H_2O$ (+ light energy) ---------- $C_6H_{12}O_6 + 6O_2$
5. Chloroplast
6. Disc shaped organelle between 2-10 micrometers with a double membrane surrounding a fluid filled matrix called the stroma. They contain structures called granum, which are made of stacks of thylakoids, which are flattened sacks where the light dependent reaction occurs.
7. Stroma is where the light independent reaction takes place, and the granum is where the light dependent reaction takes place.
8.

Adaptation	Benefit
Inner membrane	Controls what products enter and leave the chloroplast, either directly through the membrane or through carrier and channel proteins.
Many Grana	This ensures that there is a large surface area for the light dependent reaction to take place. It allows enough space for many photsystems and electron transport chains.
Photosynthetic pigments	The different types are arranged in a photosystem to allow maximum absorption of light.
Embedded proteins	These keeps the photosystems held in place.
Fluid filled stroma	Enzymes are present within the fluid which help to catalyse the reactions that take place there.

9. A molecule that allows light harvesting and provide energy to raise electrons to higher energy levels.
10. They are not specifically involved in the light dependent reaction, they do not contain porphyrin and they absorb blue light. They absorb light that chlorophyll does not and pass the energy to the bottom of the photosystem.
11. Chlorophyll A and Chlorophyll B.
12. Thylakoid membrane.
13. Source of H ions and electrons, photolysis.
14. Used for aerobic respiration, or diffuses out of leaf.
15. Hydrogen ions used in chemiosmosis, electrons to replace those lost by oxidised chlorophyll.
16. NADP.
17. To accept H ions and be used in light independent reaction to reduce carbon dioxide and produce organic molecules.
18. Making of ATP from ADP and Pi in the presence of light.
19. Molecule that transfers electrons.
20. Chemicals that accept electrons from another compound.
21. Intergranal lamella.

22. Provides energy to excite electrons.

23. They are excited to higher energy levels and accepted by electron acceptors and passed along an electron carrier system.

24. To pump H protons against a proton gradient from stroma to thylakoid space.

25. Diffusion of H protons down an electrochemical gradient from thylakoid space into the stroma, through a stalked particle which contains ATP synthase and facilitates the reaction between ADP and Pi to form ATP.

26. Cyclic uses only PS1 and non-cyclic uses both. Cyclic only prides ATP, non-cyclic produces ATP and reduced NADP and oxygen.

27. Cyclic 1 non cyclic 1 & 2

28. Reduces NADP with the addition of H.

29. Carbon dioxide is fixed and used to produce complex organic molecules

30. Stroma

31. Calvin cycle

32. Source of carbon

33. 2 ATP splits to provide a phosphate for GP and TP, Reduced NADP provides the H to reduce GP.

34. RuBP combines with CO2 in prince of rubisco to form GP.

35. Start with 5 carbon, CO2 added, which forms 2 x 3 carbon compounds (GP) , 5 carbon is reformed by 5 TP (15 C) to 3 x RuBP (15 C) (1TP used to make sugars)

36. Amino acids and fatty acids

37. Hexose sugars and glycerol

38. Combining glycerol from TP and fatty acids from GP

39. The phosphorylation of TP

40. Starch, cellulose, Sucrose and fructose

41. Ribs lose bisphosphate (RuBP), Glycerate-3-phosphate (GP), Rubisco, Triose phosphate (TP)

42. A chemical or factor required for a metabolic process in its lowest form preventing the metabolic process from producing its products efficiently.

43. Carbon dioxide concentration, water availability, light intensity, temperature.

44. Due to the increase of photosynthesis more carbon dioxide has been taken up by plants and more oxygen released so the Earth's atmospheric carbon dioxide levels have reduced.

45. By phytoplankton in the oceans.

46. The rate of photosynthesis is directly proportional to the light intensity.

47. Carbon dioxide can enter the leaves as the light causes the stomata to open, causes the phtolysis of water molecules and excites electrons in photosystems.

48. Enzymes are involved within the chemical processes of photosynthesis. Enzyme activity is directly linked to temperature as they require some heat to activate the enzymes, but if too much heat is applied then ezymes can denature and no longer facilitate chemical reactions.

49. It collects oxygen during photosynthesis and therefore measure photosynthesis rate.

50. Oxygen concentration, carbon dioxide concentration, mass of plants.

51. Placing Elodea in a test tube, collecting the gas produced through a capillary tube connected to a syringe. Changing light intensity to measure the differences in the production of the gas.

52. Same test as light intensity, however keeping light intensity the same and changing the temperature using a water bath, or adding sodium hydrogencarbonate to vary carbon dioxide levels.
53. Excite more electrons and increase the products of the reaction.
54. More NADP and ATP to convert TP into glucose and other products.
55. Less electron excitation and less NADP and ATP.
56. The stomatal opening.
57. Greater production of products such as glucose and fatty acids.
58. When stomata are open more carbon dioxide diffuses into the leaf and vice versa.
59. Less products of the light independent reaction.
60. More enzymes involved.
61. When the temperature rises above 25 degrees.
62. Reduces the overall rate of photosynthesis.
63. Damages proteins, loss of water, and stomatal closure.

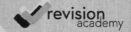

Respiration

1. The ability to do work.
2. Because it is a high energy intermediate compound and releases 30.6kJ of energy per mole.
3. Anabolic – large molecules synthesised from smaller ones. Catabolic – larger molecules are hydrolysed to produce smaller molecules.
4. Adenosine, which is a ribose sugar and adenine plus three phosphate groups
5. Energy is available in small, manageable amounts that does not damage the cell or get wasted.
6. Active transport, secretion, endocytosis, synthesis of larger molecules, DNA replication, organelle synthesis, movement, chemical activation.
7. A molecule that is organic and does not contain a protein that binds to the active site of an enzyme to aid enzyme function.
8. Glycolysis – glucose is converted to pyruvate, Link reaction – pyruvate converted to acetate, krebs cycle – reduced NAD & FAD produced along with ATP, oxidative phos phorylation – chemiosmosis produces lots of ATP.
9. Transport of hydrogen.
10. A coenzyme accepts hydrogen in glycolysis and the krebs cycle and transports it to the inner mitochondrial membrane where it releases it.
11. Bonding of adenosine, pantothenic acid, cysteine and 3 phosphate groups.
12. NAD in respiration, NADP in photosynthesis.
13. Occurs in the cytoplasm where a glucose molecule gets broken down into two molecules of pyruvate, it occurs in both aerobic and anaerobic respiration.
14. When large molecules are broken down into smaller molecules with the addition of water.
15. 2 ATP molecules are hydrolysed, the phosphate groups from the ATP attach to glucose and then fructose.
16. Providing phosphate to the sugars and energy to activate the hexose sugar.
17. Hexose bisphosphate splits into two molecules of triose phosphate.
18. 2 ATP and 2 reduced NAD.
19. Through dehydrogenation by dehydrogenase enzymes.
20. Break down of triose phosphate - 2 ATP
    Break down of triose phosphate - 2 reduced NAD
    Intermediate compound - 2 x Pyruvate   2 ATP
21. Where do the following occur:
    Where ATP is made – cytoplasm in glycolysis, across inner mitochondrial membrane in oxidative phosphorylation, mitochondrial matrix in krebs cycle.
    Where ATP is used – cytoplasm in glycolysis, other metabolic processes.
    Where NAD is reduced – cytoplasm in glycolysis, mitochondrial matrix in krebs cycle.
22. Site of aerobic respiration.
23. Rod shaped organelles that contain an envelop folded structure which hold stalked particles for ATP synthesis.
24. ATP demand.

25. How are the following adapted to their functions:
- Matrix – contains enzymes, coenzymes, mitochondrial ribosomes and DNA, and oxoaloacetate.
- Outer membrane – contains proteins to allow the passage of molecules.
- Inner membrane – folded to increase surface area, contains electron carriers.
26. They allow protons to pass through them.
27. Take hydrogen back into the mitochondrial matrix.
28. The head rotates, ADP and Pi form ATP, the ATP is then released.
29. Pyruvate is converted to acetate.
30. Coenzyme A.
31. Acetate, reduced NAD, carbon dioxide and coenzyme A.
32. Travels to the inner mitochondrial membrane.
33. Acetate combines with oxoaloacetate to form citrate, this compound gets decarboxylated twice and dehydrogenated 4 times.
34. Transport acetate into the mitochondrial matrix.
35. Accepts the acetate from the link reaction.
36. Decarboxylation and dehydrogenation.
37. Hydrogen acceptor.
38. More decarboxylation and dehydrogenation of 5 carbon compound.
39. Further dehydrogenation.
40. Draw a table to show the products made from the link reaction and Krebs cycle.

Link Reaction	Krebs Cycle
Acetate	Reduced NAD
Coenzyme A	Carbon dioxide
Reduced NAD	Reduced FAD
Carbon dioxide	ATP

41. It removes carbon from pyruvate.
42. Where ATP is formed from ADP and Pi in the presence of oxygen.
43. Inner mitochondrial membrane.
44. Series of electron acceptors that drive chemiosmosis.
45. Provide hydrogen for chemiosmosis.
46. Driven by energy provided by electrons.
47. Proton gradient.
48. Accepted by electron carriers in the inner mitochondrial membrane.
49. Final electron acceptor.
50. Oxidative phosphorylation is the formation of ATP and chemiosmosis is the build up of a proton gradient.
51. Water and ATP.
52. Leakage, some ATP is used to move pyruvate and reduced NAD form glycolysis.
53. ATP is not used to pump the protons.
54. Mitchell aimed to prove that energy from electrons was used to pump protons across the inner mitochondrial membrane to provide a proton gradient.
55. Investigations have been carried out to see if there are changes in potential difference across the membrane, and identifying the enzymes present within the system.
56. Formation of ATP in the absence of oxygen.
57. NAD and Lactate (lactic acid).

58. NAD is used to form lactate and ethanol and is regenerated to continue substrate level phosphorylation in anaerobic respiration. Within aerobic respiration the NAD carries the hydrogen to the electron transport chain in order for oxidative phosphorylation to take place.
59. Decreases pH, which affects enzyme activity and fatigues muscles.
60. Acid buffer.
61. Lactate dehydrogenase.
62. Transported to the liver and converted to pyruvate.
63. Carbon dioxide and ethanol produced.
64. Reduces ethanol to ethanol.
65. Pyruvate decarboxylase and ethanol dehydrogenase.
66. Carbon dioxide and ethanol.
67. A substance that is organic and that can be used in respiration.
68. The molecular mass in grams of a substance.
69. Lipids.
70. Hydrogen.
71. 1 mol of glucose should produce 94 mol ATP.
72. Remaining energy is released as heat to maintain body temperature.
73. Deamination in the liver.
74. In periods of starvation.
75. Some enter krebs directly, other at the link reaction.
76. More hydrogen atoms.
77. Glycerol and fatty acids.
78. Describe the process that fatty acids go through in order to produce energy.
79. Lipids.

revision academy

Linked Topic Question Answers

1. **During starvation certain respiratory substrates are no longer available, explain how the body is still able to respire.**
   Uses protein as a respiratory substrate
   Protein is hydrolysed into its amino acids
   Amino acids are deaminated into their carbohydrate and ammonia within the liver
   The ammonia is then converted to urea
   The urea is excreted through the kidneys
   The carbohydrate can then be used in respiration to produce energy
   Some protein gets converted directly to pyruvate and can then enter the link reaction
   Some can enter straight into the krebs cycle

2. **What role does the ornithine cycle play in respiration?**
   Utilises protein as a respiratory substrate
   Protein is hydrolysed into its amino acids
   Amino acids are deaminated into their carbohydrate and ammonia within the liver
   The ammonia is then converted to urea
   The urea is excreted through the kidneys
   The carbohydrate can then be used in respiration to produce energy
   Some protein gets converted directly to pyruvate and can then enter the link reaction
   Some can enter straight into the krebs cycle

3. **How are excretion and respiration linked?**
   Protein is hydrolysed into its amino acids
   Amino acids are deaminated into their carbohydrate and ammonia within the liver
   The ammonia is then converted to urea
   The urea is excreted through the kidneys
   The carbohydrate can then be used in respiration to produce energy
   Some protein gets converted directly to pyruvate and can then enter the link reaction
   Some can enter straight into the krebs cycle

4. **During periods of dehydration, water is retained by the body using the hormone ADH. How does this hormone allow the reabsorption of water?**
   Chemoreceptors within the hypothalamus of the brain.
   When water potential of blood is low, water moves out of chemoreceptors
   Chemoreceptors shrivel
   This sends a signal to the neurosecretaory cells
   Neurosecretory cells manufacture ADH and store it in posterior pituitary gland, waiting for the signal
   The signal causes neurosecretary cells to release ADH into the blood
   ADH travels in the blood to the collecting duct
   ADH acts on collecting duct cells by increasing their permeability to water
   Water is reabsorbed into the blood and water potential of blood increases.

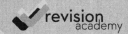

5. **What is the significance of cell membranes is hormone secretion?**
   Protein based hormones cannot enter cells
   Cell surface membranes contain receptors
   The receptors are specific to the hormonal proteins
   These receptors are attached to other substances within the cell
   Once activated by the hormone receptor complex other changes in the cell may occur
   These can be long lasting as hormones can stay in the blood permanently as they can be constantly manufactured.

6. **How do insulin and glycogen regulate blood glucose levels?**
   Insulin reduces glucose levels
   Glucagon increases glucose levels
   Detected by alpha and beta cells in the pancreas
   When blood glucose is too high
   The carbohydrate can then be used in respiration to produce energy
   Some protein gets converted directly to pyruvate and can then enter the link reaction
   Some can enter straight into the krebs cycle

7. **Describe the role of respiration in temperature regulation of endotherms.**
   Endotherms maintain internal temperature despite external conditions
   Respiration of respiratory substrates produces ATP
   The rate of respiration within the liver can be altered
   If more heat is required, more respiration in the liver takes place
   More heat is released
   A significant amount of energy from food is used to maintain body temperature

8. **How may a dramatically reduced salt intake effect the sensory system?**
   Neurones contain sodium/potassium pumps and channels
   The electrical impulse that travels along neurons is in the form of sodium and potassium ions
   Salt contains sodium
   With a decreased amount of salt there will be less sodium to be used in nervous impulses
   Less sodium means there will be less available to be actively pumped into neurones
   Therefore a resting potential will not be able to be maintained as a polarized membrane
   Depolarisation will continue but at a lower rate
   A threshold potential may not be reached
   Not action potentials will be initiated
   This could lead to loss of senses and possible eventual death

9. **A person consumes extremely high amounts of salt constantly and begins to feel numbness, loss of taste, hearing and other senses. Why is the person experiencing these symptoms?**

Neurones contain sodium/potassium pumps and channels

The electrical impulse that travels along neurons is in the form of sodium and potassium ions

Salt contains sodium

Increased amount of sodium will make it difficult to maintain a resting potential initially

The neurone and surrounding tissue may become flooded with sodium ions

Therefore an electrochemical gradient cannot be achieved

Neither resting potential or local currents can be created

Therefore an action potential cannot be generated even when a stimulus is present

With no action potentials the person will be unable to send stimuli messages to the brain to be interpreted

Hence loss of senses

10. **MS is a disease which effects the myelin sheath of neurons What symptoms may a sufferer experience and why?**

The myelin sheath aims to insulate the electrical impulse

Saltatory conduction occurs between the nodes of ranvier

When insulated the impulses can travel faster

Myelination is generally found in peripheral nerves

With damaged schwann cells the impulse will not be insulated

Therefore the impulse may travel slower or be lost and not reach the CNS

MS suffer with feeling in skin, eyesight, and some are unable to walk

11. **Cyanide is a respiratory inhibitor, which declines and ultimately ceases the production of ATP. How might this effect nervous transmission?**

ATP is required for maintaining the resting potential of the neurone

It is used to actively pump sodium ions out of the cell to produce an electrochemical gradient

With no ATP, no sodium is pumped across the membrane

Sodium ions will diffuse across the membrane initially stimulating an action potential if the threshold potential is reached

Ultimately the sodium ions will reach equilibrium and the electrochemical gradient across the membrane will be lost

No further action potentials will be intitiated

12. **Why is it important for mitochondria to be present among neurons?**

Mitochondria are the site of cellular respiration and production of ATP

ATP is required for maintaining the resting potential of the neurone

It is used to actively pump sodium ions out of the cell to produce an electrochemical gradient

This provides a starting point for action potentials to be initiated when the resting potential is disrupted and the membrane become depolarised

Therefore mitochondria are essential for provided the necessary ATP

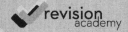

13. **How might a calcium deficiency affect the sensory system?**
Calcium ions are involved in synaptic transmission
An action potential arrives at the synaptic knob
This causes the voltage-gated calcium channels to open
Calcium ions diffuse into the synaptic knob
This causes synaptic vesicle to move towards and fuse with the pre-synaptic membrane
A neurotransmitter is released from the vesicles and diffuses across the synaptic cleft
They then bind to receptors on the post synaptic membrane initiating an action potential within that neurone
If calcium ions were deficient this would prevent the release of neurotransmitters into the synaptic cleft
This would prevent the continuation of the action potential

14. **A certain gas is known to non-competitively inhibit acetylcholinesterase, what would be the effect of inhalation of this gas?**
Acetylcholinesterase is the enzyme used to break down the neurotransmitter acetylcholine
When acetyl choline is released into the synaptic cleft it binds to receptors on the post synaptic membrane
Usually acetylcholinesterase will then break down the acetylcholine to prevent it further initiated action potentials in the post synaptic neurone
A non-competitive inhibitor will bind to a site away from the active site on acetylcholinesterase
This will change the active site of acetylcholinesterase
Acetylcholine will not longer be able to form an enzyme-substrate complex
Acetylcholine will not be broken down
More action potentials will be initiated in the post synaptic neurone

15. **Some substances are able to bind to the postsynaptic membrane in neurons, blocking them from receiving any further substrates. What effect will this have on action potential transmission?**
When a neurotransmitter is released into the synaptic cleft it binds to receptors on the post synaptic membrane
This will initiate an action potential in the post synaptic neurone
If the binding sites of the post synaptic membrane are blocked the neurotransmitters will not be able to bind with the post synaptic membrane
This will prevent action potentials being initiated within the post synaptic neurone
The message will cease to be sent along the nerves

16. **A person constantly consuming a poor diet, particularly high in sugars, begins to feel diabetic symptoms. Explain the effect of poor diet on insulin production.**

Insulin is used to convert excess glucose to glycogen

The insulin makes liver and muscles cells more permeable to glucose

Once in these cells the glucose is converted to glycogen or fat and stored for later use

With continual excess glucose in the blood stream, more and more will be stored as fat

Causing the individual to become overweight

Constant stimulus by insulin, the liver and muscle cells and become somewhat 'insuli-resistant'

Therefore insulin is less effective at reducing the blood glucose level

In some cases insulin levels then increase and cause other problems such as infertility in females

17. **During an accident a patient suffers a severed vagus nerve. What effects will this have on the body and how could doctors remedy the situation?**

The vagus nerve connects the cardiovascular centre within the brain to the sinoatrial node within the heart

The cardiovascular centre contains receptors that monitor movement, carbon dioxide levels, adrenaline, blood pressure

If increases are detected an impulse is sent down the vagus nerve to increase heart rate and therefore respiration rate

If this nerve is severed, respiration rate cannot be regulated in response to different stimuli

18. **Describe the cascading effects of exercise on the body's systems.**

Muscles contain stretch receptors which are activated when the muscles move

These send impulses to the cardiovascular center which in turn sends the signal to the heart via the vagus nerve

This will increase heart rate

More carbon dioxide is produced as the muscles are using up more oxygen to produce ATP for muscle contraction

Chemoreceptors detect the change in pH of the blood when carbon dioxide forms carbonic acid in the blood

This signal is sent to the cardiovascular center and then onto the heart to increase heart rate also

19. **Some doctors claim that a small, regular intake of alcohol can be beneficial. Explain why this may be so.**

The liver is responsible for the detoxification of alcohol

The ethanol is dehydrogenated by the enzyme ethanol dehydrogenase

The product being ethanal

This ethanal is further dehydrogenated by ethanal dehydrogenase into ethanoic acid

This combines with coenzyme A to form acetyl coenzyme A

This can be used at the base of the link reaction to drop off the acetate to be used in the krebs cycle

This can therefore aid respiration rate

Acknowledgements

Special thanks to Ben Richardson for proof reading and checking through the workbook.

We would also like to thank Amanda Mann & Clare Colwood for help with data input.

The author and the publisher would like to thank the following for permission to reproduce the following images:

Cover: Mopic/shutterstock

p.132: Knorre/shutterstock, dream designs/shutterstock, Creations/shutterstock

Also in this series.......

ISBN 978-1-910060-02-5

ISBN 978-1-910060-03-2

ISBN 978-1-910060-01-8

Publishing

www.synthus.co.uk

Printed in Great Britain
by Amazon.co.uk, Ltd.,
Marston Gate.